KEEPING GARDEN CHICKENS

EVERYTHING YOU NEED TO KNOW ABOUT KEEPING CHICKENS IN THE UK

A M STOKER

Keeping garden chickens everything you need to know about keeping chickens in the UK

Copyright © 2024 by A M STOKER

All rights reserved. No part of this book may be reproduced, distributed, or transmitted in any form or by any means, including photocopying, recording, or other electronic or mechanical methods, without the prior written permission of the publisher, except in the case of brief quotations embodied in critical reviews and certain other noncommercial uses permitted by copyright law.

Published by: A M STOKER

Illustrated by: A M STOKER

Country: UNITED KINGDOM
Published date: 2024

First Edition: 2024

This book and its entirety is protected by UK copyright © law.

Content of this book

Page 10.
So let's get to it.

Page 11.
Are chickens the pets for you?
Legally, can you keep chickens in your back garden?
Consider your neighbours?

Page 12
Do you have enough space to keep your fluffy girls?
Do you have the time to be a poultry keeper?

Page 13.
The question your partner keeps asking you. How much is this going to cost me? Can we afford it?
I've put together a basic startup list.

Page 14.
So with all that in mind, if chickens are really for you and your family, roll up your sleeves and please read on. We will now get into The nitty gritty parts.

Page 15.
Housing your little chicken.
Building your own coop.

Page 16 17 18 19 20.
Building your own coop.

Page 21.
All your other essentials.
Food dispenser.

Page 22.
Food.

Page 23.
The good stuff.

Page 24.
Foods to avoid.

Page 25.
Foods to avoid.
Water drinker.

Page 26.
Water drinker.

Page 27.
Bedding.

Page 28.
Bedding.
Material for the run.

Page 29.
Cleaning equipment.
List of cleaning equipment.

Page 30.
List of cleaning equipment.
First aid kit.

Page 31.
First aid kit.

Page 32.
Finding the right poultry supplier.
Qualities of a good supplier.

Page 33 34.
Important questions to ask the supplier.
Point of lay.
Have they had their wing clipped?
Worming.
Have they had their vaccinations?

Page 35 36 37 38.
So what breeds should you go for?
Warren.
Blue bell.
Light sussex.

Page 39.
So the time has come to pick your new girls.
How to pick up a chicken.

Page 40.
Your checklist of things to look out for.
Posture and walking.
Eyes and nose.
Breathing.

Page 41.
The comb.
The ladies legs.
Feathers.

The fluffy butt end.

Page 42.
Time to take your hens home.
Now you're home. What do you do next?

Page 43 44.
Pecking order.

Page 45.
Keeping your girls entertained.

Page 46.
Housekeeping.
Poultry shield.
Daily clean.

Page 47.
Deep clean.

Page 48 49.
Creatures of the night, red mite.
Red mite on your girls.

Page 50.
Lice.

Page 51.
Worms.

Page 52 53.
The fluffy butt nuggets.
Normal poop.

Page 54.
Cecal poop.

Page 55.
Intestinal lining poop.
Bloody poop.

Page 56 57.
Bloody poop.

Page 58 59.
Diarrhoea.
Green poop.

Page 60
Worms in poo.

Page 61.
Health and well-being.

Page 62.
Prolapse.
Why a prolapse occurs.

Page 63.
Preventing a prolapse.
How to treat a prolapse.

Page 64 65.
How to treat a prolapse.

Page 66.
Egg bound chicken.
Causes of egg binding.

Page 67 68.

Signs that your girl is egg bound.
How to treat an egg bound chicken.
How to avoid egg binding.

Page 69 70.
Vent gleet.

Page 71.
Respiratory disease.

Page 72 73.

Signs of respiratory disease.
Preventing respiratory disease.

Page 74.
Impacted crop.

Page 75.

How to prevent a Impacted crop.
Signs of Impacted crop.

Page 76.
Treating a Impacted crop.

Page 77.
Treating a sour crop.

Page 78.
Bald bottom chickens.
How to care for your girls when they are moulting.

Page 79.
Scabby Scaly legs.

Page 80.
Eye issues.

Page 81 82.
Your girls and other animals.

Page 83 84.
Introducing new chickens.
The right way to integrate chickens in three simple steps.

Page 85 86.
Producing the goods, eggs.

Page 87.
Maintaining a regular supply of healthy eggs.

Page 88.
What to do with your yummy eggs.
Not laying eggs.

Page 89 90.
Soft shell eggs.
Eating the goods.
Preventing egg eating.
If they are eating through your enterprise.

Page 91 92.
A final note.
Thank you page.

Page 93.
Books published by A M STOKER.

So let's get to it

Back garden poultry (Chicken's) in the UK has always been a popular choice for families.

Especially during WW2. It was very cheap and easy for people to keep chickens in their back gardens. As they would feed the chickens scraps from the dinner table and in return a regular supply of eggs to feed the entire family.

The number of back garden chicken keepers is becoming more and more popular and I can understand why, the quality of eggs are amazing.

I personally believe keeping chickens in your back garden is great for the whole family and surprisingly, they are particularly good pets. I have wasted countless hours just sitting there watching them. They are surprisingly entertaining and funny. Well, I believe they're more entertaining than some of the reality programs my wife watches on TV, but don't tell her I said that.

So this book is about my personal experience and knowledge of everything I have learnt over the years as a poultry keeper. I'm not suggesting I am an expert. I am a back garden poultry keeper in the UK, that has experience and information I believe could help new back garden chicken keepers like yourself. I also wanted to write a book for you that had all of the information you would need to know, in one place. And oh boy, what I'm about to share with you, I wish I knew when I first started keeping chickens.

Are chickens the pets for you?

I have a saying, if you're going to do something, do it properly or don't do it at all. I'm sure you heard it before but it's true. So with that in mind, first things first.

Legally can you keep chickens in your back garden?
Before you bring your chickens home and before the whole family becomes attached to our feathered friends. I would check to see if you are legally allowed to keep chickens in your back garden. You can do this by looking at the deeds of your house and contacting the environmental health office.
If you are allowed, be aware that you are legally required to register your chickens online, which is very easy to do so.
You can do this by going on the internet and typing - Register my chickens, then go to gov.uk page and follow the instructions. Easy as that. It is a good idea because they will let you know via email or text message if any disease outbreaks are in your area.

Consider your neighbours?
If you have a small garden that is close to other houses, I would consider the fact that your neighbours may find it a surprise to wake up to the sound of chickens singing to them early in the morning. From my experience, most chickens will make noises that we call the egg song, when they are ready to lay an egg or after they have laid an egg.
Some chickens are louder than others, so to keep your neighbours happy, I would consult them and make sure that they are happy with you keeping chickens. The last thing you need is them constantly complaining about your new chickens.
You could always bribe them with some yummy eggs. That tends to work.

Do you have enough space to keep your feathered girls?

And no, I don't mean a third bedroom in the house. Chickens will live in the garden in their own little house called a COOP, this is where they will sleep. Each chicken will need a minimum of 26 cm sq sleeping space. Attached to the coop will be what's called nest boxes; one nesting box is enough for 3 chickens.

Attached to the coop will be a RUN, an outside space. This needs to be a minimum of 2 m sq per chicken.

Do you have the time to be a poultry keeper?

Keeping chickens is really a labour of love in my opinion. Me and my family thoroughly enjoy our time with the chickens, and yes they all have names but I'll keep the names to myself. I don't need my street cred getting any lower than it already is.

So they need to be looked after 7 days a week 365 days a year, like all pets that we love.

However, don't panic, it only takes up to an hour of your time per day, unless you're like me, where you get hypnotised and sit there for hours on end.

Half an hour in the morning, half an hour in the evening.

Every morning you will need to let the girls out of the coop, top up their food and give them fresh water. As well as giving them a little checkover to make sure that they're all healthy. If you can, clean out the droppings in their coop.

And of course, most importantly collect your eggs.

In the evening you'll have to top up their food if needed and give them fresh water if needed. As well as clean out the droppings from their run. If you didn't clean the coop in the morning, then that must be done before you tuck them up in bed.

The question your partner keeps asking you, How much is this going to cost me? Can we afford it?

I'm not going to lie to you. The initial startup can be quite expensive.

However, there are ways of making it cheaper for yourself, for example, if you or anyone in your family are a dab hand at DIY then you could build your own coop and save yourself a reasonable amount of money. This is what I did and I have saved a lot of money compared to buying a coop online. However, the online coops are particularly good, especially if you are not very good at DIY. I tend to find however that they are not very big. This is why I preferred to build one myself. I found It was more suitable for me and I could get a lot more for my money.

I've put together a basic start up spending list for you.

- Per chicken £20
- Coop + run. 4-6m £300 - 1300
- Food and water dispenser 3l £26
- Food (layer pellets) 20kg £11 - 16
- Poultry tonic 500ml £18
- Bedding 20kg £16
- Cleaning fluid (poultry shield) 5l £18

Prices are in 2024, who knows what the cost will be in a few years, with the way the UK is heading. This is just a rough guide on what the initial outlay of costs for starting up will be. However, most of the items here will last you a long time and it does not necessarily have to be replenished regularly.

Food and bedding depends on the amount of birds that you have.

So with all that in mind, if chickens are really for you and your family, roll up your sleeves and please read on. We will now get into The nitty gritty parts.

Before you get over excited, rush to your local poultry farm and buy a bunch of chickens. There are a few things to prepare first.
It is best practice to have the coop built and ready for your girl's to move straight into, with their food and water ready and their bed made.
The reason for this, is that the first 48 hours are extremely stressful for them, so having everything in place reduces the amount of stress put on them.

Housing your little chickens

Ok so this is going to be the expensive part but if built properly you Shouldn't have to do it again.

Without a doubt, plastic coops are a great option. They are very easy to keep clean. Easy to build and last forever. Plastic is generally the best at lowering the chances of red mite Infestation. However, the downside to plastic coops is they can be quite pricey. But well worth looking into.

There are plenty of good websites selling wooden coops. Great if you are not up to building one yourself and you want to keep the cost down.
You will have to assemble it yourself which is reasonably easy, I would recommend painting it to protect the wood. It will last longer if you do.
Please avoid felt roofs as it's a breeding ground for red mite. We will move on to red mite later. Trust me you don't want them!
The down side to most of these types of coops is that they are small, which means normally you can only have between four and six chickens at best. They are also not very tall, so you spend most of the time on your knees whenever you have to do something inside the run.
When buying online, make sure you check the full dimension, to make sure it is large enough to house the number of chickens that you intend to keep.

Building The coop yourself
From experience I found building your own coop is much more practical, and if you shop around you can pick up all your materials

for roughly the same price as you would buying a wooden house online.

The perks of building one yourself, It can be built to the shape and size of your garden or space that you have available. You have a choice on how it looks and how it fits in with the design of your garden. You can build it to suit your practical needs. How crazy you want to go is up to you.

Building a coop and a run above head height is a game changer. As it makes it easier for you to go in and clean, change their water and top up their food or anything else that you need to do.

Make sure you have easy access to their house (COOP) as this will make cleaning out their bedding as easy as possible.

Inside the coop should be perches, 4 to 5 cm square rounded off timber. Make them removable for easy cleaning. They should be roughly between 30 to 50 cm off the ground.

It is very important to make sure that the coop has adjustable ventilation holes at the top of the house. It's important so you can control the airflow and the temperature of the house. The more airflow the better. Don't close the vents until the temperature drops below zero.

Attached to the coop is a sliding door. This allows your chickens to go too and from the coop, to the run. I recommend that this is 41 cm high by 36 cm wide.

I prefer to have the nest boxes attached to the outside of the coop, as it makes it easier for you when you are collecting your yummy eggs. The boxes need to be roughly 30 to 50 cm off the ground. It also needs to be as dark as possible, this is because the girls find it more comfortable when laying. It also prevents them pecking at the eggs.

Having a waterproof roof for the coop and run is a must. You will thank me for that, when the rain is coming down thick and fast and in the UK! This happens a lot.

Chickens don't like the rain either so that will keep them happy.
Do not use felt roofing whatever you do. Just trust me.

The run is built from timber and lined with welded mesh. This is so they can see every movement that you make and shout at you when they see that you're not showing them enough attention. Welded mesh is also much stronger and protects your chickens from predators.

I would advise building a welded mesh divide with a door, in the run. Then you have an isolation pen if any of your chickens get ill and there will come a time when they do.
Or, if you're wanting to integrate new chickens to your current flock. It should be built with a door. So you and your chickens can still use the entire run when it is not needed as a pen, by simply keeping the door open.

Make sure you beef up the security and make it as predator proof as possible. Foxes are the main predator that you have to worry about breaking in and they are sneaky little devils.
So strong locks and good quality welded mesh for the run is a must. Do not use chicken wire as it is not strong enough and foxes can easily break through it.
If you plan on putting the coop on soil, there's a few ways to protect your girls from foxes getting in:

. If your run is on soil, then one of the easiest things you can do to protect your chickens, is to place concrete paving slabs around the entire perimeter of your chicken run. I would recommend 1 ½ to 2 ft square slabs.

. Around the perimeter dig 1ft down, line it with welded mesh or fill it with poured concrete.

. Around The perimeter you can lay flat the welded mesh and attach it to the coop by nailing it to the bottom. Then cover with soil or a flower border and compact it down. This will stop the foxes from digging a hole as the fox cannot scratch through the mesh.

I recommend 2 ft from the coop out. Foxes are not smart enough to realise that if they went 2 ft back they could dig under and through. They will generally dig as close to the perimeter as possible.

Most importantly, you can build the coop and run to the correct size to the amount of chickens that you desire.

Coop.	Per chicken.	26cm sq.
Run.	Per chicken.	2m sq.
Nest box.	1 box to 3 chickens.	35cm sq.

There are plenty of detailed designs and plans online that will help you build your coop. There are also many great videos on YouTube of people building their own coops from start to finish, which will give you a good idea of how to build it and what will suit you.

This picture is a particularly good design, it covers all the needs required for a good home for your fluffy girls.

When building your own coop you've got to think about red mite. They like to breed in every nook and cranny. The best way to get over This is to use GRIPFILL. Seal every crack you can, If you can get a thin piece of card in the cracks or joints then red mite can live in there. Don't use silicone as poultry shield deactivates the silicone and it will peel off.

Use smooth sanded timber, this will also help prevent red mite as it's more difficult for them to bury themselves in and lay their eggs, whereas rough timber is much easier for red mites to do so and we don't want that.

If you use a weathershield paint, apply two to three thick coats. This will prevent red mite from bedding into the wood and it'll be much easier for you to spray and wipe down when cleaning.

I'm not suggesting that you have to build Cluckingham Palace but the better you build your coop, the easier the experience will be for you and your girls.

Like I said, if you're going to do something, do it properly or don't do it at all.

All your other essentials needed

Food dispenser

There are many different types of food dispensers and drinkers. I believe it's a personal choice on what you prefer to use. I prefer to use a good old-fashioned plastic dispenser and drinker, it's cheap, easy to keep clean and reliable. And they come in many different sizes.

My top tip for you is, for however many chickens that you have, make sure the dispenser holds enough food for a minimum of 3 days. It's also worth raising it off the floor to avoid unwanted items in their food.

Food

The most important part, without doubt, of a chicken's diet is layer pellets, this is what they should be fed 90% of the time. This is because a good quality layer pellet will have all the nutrition that your girls need.

Whatever you decide to feed your chickens, organic, GM Free it's totally your own choice. When buying layer pellets just make sure it contains a minimum of 16% protein.

One of the best all-round layer pellets that I use, with all the nutritional requirements and value for money is (Marriages Farmyard Layers Pellets 20kg).

If you want those eggs to keep popping out, then don't be shy. Make sure enough food is accessible throughout the day.

Each chicken will normally need to consume 120g of layer pellets per day depending on how greedy they want to be.

You are welcome to feed your flock treats. I recommend doing this only in the evening, 3 hours before they go to bed and remember only 10% of their diet can be treats.

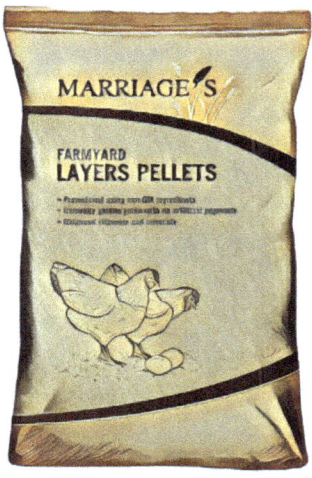

The good stuff

Green vegetables are a great source of vitamins and minerals.

Soft fruit contains a good amount of vitamins, however I recommend just a little bit now and then, or your nice clean patio may look like a painting of modern art.

Banana, great for potassium but only once every other week and a small amount. I've heard it's great for hangovers as well.

I need to make you aware that it is against the law in the UK to feed poultry kitchen scraps, meat, or anything from your kitchen as a matter of fact. As silly as it might sound, it is for good reason. It is to avoid cross-contamination, viruses and disease.

If you grow your own vegetables in your garden and prepare them in your garden, away from the kitchen then you are not breaking the law, as far as I'm aware.

I found the safest and easiest way to treat my girls is to feed them mixed corn and don't they love it! Please only feed a small amount. If fed too much it can lead to problems.

. They can become overweight, and fat chickens don't lay as many eggs.

. Too much grain can affect their digestive system and cause a Impacted crop.

It is important that you feed your hens a little chicken grit. This acts as teeth and breaks down the food in their digestive system. Crushed oyster shells help maintain strong eggs by providing calcium. It's best to put a small amount in their feed.

A great tip for you, when you treat your girls with mixed corn, put the corn in a tray, shake it and repeat "come come". By doing this you will teach your girls to follow you which will make it so much easier when moving them around.

Food to avoid

. Chocolate, although I can't get enough of the stuff, chocolate as it is for most animals, is poisonous for chickens.

. Potatoes, raw or green potatoes and potato peels Believe it or not, are toxic to chickens, cooking potatoes, breaks down the toxic compound and is ok to feed chickens. My advice is just to avoid it.

. Avocados are also toxic and can stop the heart.

. Apple seeds, The seeds contain cyanide which can kill your chickens. Any other part of the apple is fine, so when giving them apples make sure they are seed free.

. Onions, again just avoid it. It's safer.

. Rhubarb can cause severe liver damage and death.

. Meat and all meat products. It is against the law to feed any meat or meat products to your chickens in the UK and for good reason. Meat comes from all round the world and does pose a good chance of bacteria and viruses. Even when cooked, there is a very high possibility that viruses could still survive, this will cause big problems for your chickens and for everybody else.

. Mealworms, most mealworms are imported and could have come into contact with, or been fed, animal protein which could then potentially pass on disease. In 2014 DEFRA banned the feeding of mealworms to chickens in the UK. Imported mealworms which have not been inspected and certified means there is no way of knowing if they have been ingesting animal proteins or not. So in other words, it's just not worth it.

Poisonous plants for poultry

Bull Nettle	Horse Chestnuts
Bryony	Hydrangea
Bracken	Hyacinth
Bloodroot	Ivy
Cocklebur	Lily of the valley
Careless Weed	Laburnum
Castor Bean	Lantana
Curly Dock	Lamb's Quarters
Delphinium	Nightshade
Daffodil	Rhododendron
Elderberry Fern	Rhubarb
Foxglove	St. Johns Wort
Ground Ivy	Tulip
Hemlock	Water Hemlock
Horse Radish	Yew

Last of all, nicotine can easily kill chickens, so please keep all tobacco and cigarettes away from your fluffy girls.

Water drinker

As you can imagine, this is pretty straightforward. I recommend that you change the water everyday and clean the drinker every week with poultry shield, rinse thoroughly with warm water before refilling with freshwater for them to drink.

In the winter Apple cider vinegar with garlic is great. This will keep their tummies warm and it is really good for their immune system. Follow the instructions on the bottle.

Poultry tonic is great for chickens when they come into moult, It gives them all their vitamins and minerals that they would need to recover. It's also a good pick me up for when your girls are feeling unwell.
In the first week that you get your chicken's home, it is best to give it to them for that first week. Follow the instructions on the bottle. When your chickens become unwell, some medications are added to the water. They must drink the entire content of the medication before you change it for freshwater.

Surprisingly hens drink a lot of water, especially in hot weather. It also helps with their digestion. If hens have a regular supply of clean water they will lay more consistently and produce healthier eggs, as about 75% of egg is water. When buying a drinker, bear in mind that each chicken will drink between 300 ml to 500 ml per day.

Bedding

So when I talk about bedding, I'm not talking about the super king bed in the deluxe suite. Hens rarely lay down, they tend to roost (sleep on perches). So why do chickens need bedding? They need it for insulation during cold times and they need a material that serves as litter for their droppings.

The qualities of good litter.
. Good moisture absorption/release.
. Keeps odours down.
. Dries out droppings.
. Doesn't decompose.
. Provides warmth during the winter.

Straw

Is a popular chicken bedding, it's easy to get hold of, it's good for insulation and your girls will enjoy scratching around in it. However in my experience it struggles to release moisture and it holds on to the bad smells a bit more than the rest of the types of bedding.

Wood shavings

It's very absorbent so it will get rid of excess moisture in the coop. It's not very good for insulation. Your girls will need something that will keep them warm in the winter months. However it is reasonably cheap if you're trying to keep the cost down and it's easy to get hold of.

Hay

The difference between straw and hay is, hay is a crop, whereas straw is a by-product of grain crops. Hay is often used as a feed for livestock during the winter when fresh grass is not available. So for this reason, I do NOT recommend using hay as bedding for your girls. They will eat it, which can lead to an Impacted crop. Hay can also develop mould spores due to its high absorbency rate which can cause even more problems like respiratory disease.

Hemp

This is what I use and what I would recommend to you. After trying all the other options of bedding, hemp has all the qualities needed for a great litter. Which is the whole point of bedding. It's easy to clean out the droppings without wasting any hemp, you just need a cat litter scoop. Yes, it is a little more expensive and not as easy to get your hands on. However it lasts a lot longer and goes a long way. In this instance you pay for what you get.

Material for the run

What you choose to use on the ground in the run is a personal choice.

- *Hardwood chip* Hens love it, it stops the ground getting too muddy and it can be pressure washed.

- *Smooth stones* Hygienic and easy to pressure wash. However, it has been said that a lot of chickens don't like to walk on them.

- *Sand or gravel* This is very good as it acts as a litter tray and you can use a poop scoop to clean out your girl's droppings, just like you would with a cat litter tray.

- *Concrete* Very hygienic as it can be pressure washed

- *Grass* The girls will love it. However, if you leave the run in the same place, then it won't take long for the grass to turn into mud.

- *Soil* The girls love it but if you let them, they will dig down to China.

Cleaning equipment

So, this will be some of the most important stuff that you will need and use on a regular basis.

Keeping the chicken area clean is very important for their health and well-being. A dirty coop can lead to some poorly chickens.

Hens spend most of their time shoving their faces in the dirt foraging for bugs and insects. As you can imagine, if the ground is dirty and covered in butt nuggets, it can easily lead to your girls becoming sick.

If you're squeamish and don't like picking up poop then chickens are not for you. It's not as bad as you think though.

You get used to it. It's not like you're picking up a big pile of steaming dog poo.

There is one good thing about cleaning up the chickens' droppings, you can tell a lot about their health by their poop. There will be a fascinating section all about this further on in the book.

List of cleaning equipment

Poultry shield

Is by far one of the best products on the market for cleaning and prevention of unwanted creepy crawlies.

It comes concentrated so you will have to mix it with clean water in a spray bottle. A 5l bottle will last you a long time.

Spray bottle

For the poultry shield.

Rubber gloves

I use the 100 packs of disposable rubber gloves. Perfect for protecting your wife's acrylic nails. Nobody likes a poop finger.

Paint scraper
Perfect for anything that's proving tough to shift.

Bucket and small bin bags
To put the poop and dirty bedding in whilst you are moving around the coop.

Wet wipes. dustpan and brush. Microfiber cloths. Scrubbing brush. Poop scoop.

First aid kit

When keeping poultry in your back garden, whether you like it or not, there will come a time when your girls become sick or injured. There are a few things that you can have in your first aid kit that you will need.

. Poultry shield for killing off red mite.

. Diatomaceous earth or Diatom, (it is the same thing). This helps to prevent red Mite by spreading it around the coop.

. Cotton buds and cotton pads. You would need this for when they have a runny nose or blocked nostrils. By soaking the cotton in warm water, you can wipe their nostrils gently.

. Antiseptic spray, animal friendly antiseptic spray can be sprayed onto cotton wool pads to clean cuts and wounds.

. Vaseline is used if your girls have scaly legs.

. Poultry tonic, this gives your chickens all the vitamins and minerals that they need when they are moulting or recovering from moulting. It's also used when chickens are stressed, or if your girls are feeling a little bit unwell, this will help perk them up.

. Apple cider vinegar with garlic added to their water will maintain inner health, maintain a healthy gut, fight off bad bacterias and boost their immune system. It will help with respiratory problems and can even help stimulate egg production.

. Hibiscrub or iodine. You will need this if your girls have a prolapse and it's good for many other things.

. Rubber gloves. There may come a time where you have to push a prolapse back in.

. Syringe. If you need to deal with an Impacted crop.

. Preparation H. A cream to help with a prolapse.

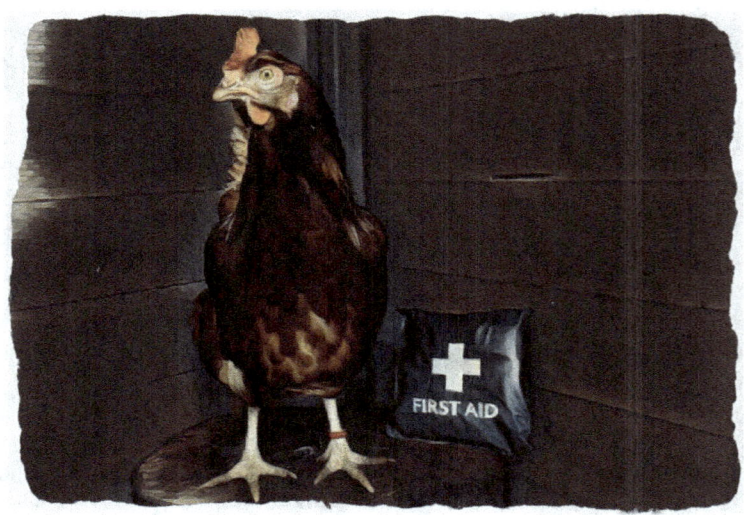

Finding the right poultry supplier

As more and more people in the UK are keeping chickens in their back garden, as well as farmers expanding their flocks. The demand is always increasing, which means you shouldn't have any problems finding a supplier to purchase your chickens.

My advice to you is, do your research on the poultry suppliers in your area. Make sure they are a well established supplier with good reviews.

Qualities of a good supplier

An ideal supplier should be willing to answer any of your questions before and after your purchase, especially if your girls become unwell. You should be able to contact the supplier for advice and possibly they may be able to assist you with basic medication, before the need to visit a vet.

A good supplier will also advise you on vets that specialise in poultry care.

A confident supplier who trusts their stock of chickens should give you a short warranty on their hens and aftercare.

A good supplier should have vaccinated all of the chickens they are selling with all the necessary vaccinations required. They should also give you a full list of vaccinations your chickens have received. You will need this if your chickens become poorly and need to visit a vet, as the vets will ask what vaccinations your hens have received.

The farmer that I use for all my poultry is always happy for me to contact her with any concerns that I may have. She is always willing to receive emails with detailed description and photos of any illnesses or injuries that my chickens have obtained. She will advise me in the best possible way that she can and sometimes provide me with medication.

Important questions to ask the supplier

Point of lay
This is the right time to purchase your chickens. 'Point of lay' is the terminology used when the age of a hen is between 16 to 20 weeks old and the start of their laying life. This doesn't mean they're going to start popping out eggs as soon as you get home.
They need time to adjust to their new surroundings and new home. They also need to get over the stress of the move. It can even depend on what type of weather and what time of year. Don't be surprised if you don't see eggs for 2 to 3 weeks.

Have they had their wing clipped?
Most suppliers would not have their chickens' wings clipped as there is not much need at a poultry farm.
However, yes it's true chickens can fly but not far. My advice to you is, if you have neighbours close by and with their own pet, I would definitely have your girl's wings clipped.
It's not as easy as you think, explaining to the whole family why there is a pile of feathers on the lawn but no chicken.
The supplier will be happy to show you how it's done. Only one wing will need to be clipped.

It's not difficult to do, he said, so you will have no problems when it needs doing again. Which will be roughly each year when the feathers grow back.

If you have a very big garden with plenty of space and no neighbours close to you, and you feel that the girls will be safe if they took a little trip over the fence, then the choice is yours whether you do or not.

It's not like you're severing their left hand from their body, You're just clipping a few feathers. It does not hurt them.

Worming

It is important to know if your girls have been wormed or not. I personally wouldn't purchase chickens without them being wormed first.

This is done with a product called Flubenvet.

Have they had their vaccinations

Most good poultry suppliers will fully vaccinate every chicken they have, as it's not worth risking the entire flock becoming sick, just to save a few quid on cutting back on vaccinations But it is still important that you ask if the girls have been fully vaccinated and for a full list of the vaccinations that they have received.

I will say that even if they've had their vaccinations, it doesn't guarantee that they are fully protected. It is still possible for them to get viruses, but you will have a lot less problems if they have been vaccinated correctly.

Ok, at this point if the alarm bells are not ringing. The supplier has happily answered all your questions above and supplied you with all the information that you need, then it is time for you to take a look at the girls.

So what breed should you go for?

There are hundreds of different breeds out there. Purebred, pedigree, you name it, I could write another book just on the breeds alone, so I'm not going to get into all of that.

What I'm going to be talking about is hybrid chickens. The development of hybrid chickens started in the 1950s because of the growing demand for eggs and meat. The idea of hybrids was to lay a regular supply of healthy eggs at the lowest cost possible.

Hybrid chickens are a cross between different pure breeds, such as the Rhode Island Red, Sussex, Plymouth Rock, Leghorn, and Maran.

Hybrid chickens are much cheaper to purchase than their fancy purebred sisters.

However, hybrid chickens are known for being strong and healthy girls. Hybrids are also great egg layers, they can lay over 300 medium to large eggs per year, plenty to feed the whole family.

Hybrid chickens are bred to be calm and peaceful, generally they are easy to look after. That makes them a great pet for the entire family.

If you're looking for a hen that is very easy to look after, produces lots of eggs and is an ideal pet for the children, then look no further than the Warren.

Warrens

In my opinion, they are the ultimate chicken. Warrens are primarily Rhode Island red strains.

If they are fed well with a balanced diet they will produce around 300 to 320 eggs per year. They will lay an average of 63 g per egg, some of my warrens will lay up to a whopping 83 g per egg. In my opinion, that's more like giving birth than laying an egg. However, I'm not complaining when I'm having my breakfast in the morning.

Warrens tend to lay eggs for 3 to 4 years and they can live to 5 years old.

Bluebell

Bluebells are a large hybrid chicken. I would say they are one of the most beautiful looking girls out there, definitely a head turner. Bluebells usually are a cross between Blue Marans and Rhode Island Red. They have a blue plumage that ranges from pale to darker in tone. The bluebell is another very calm and relaxed hen. Friendly and confident. I think they make lovely pets for the family. They are also good in mixed flocks. These girls will lay up to 260 large brown or pinkish eggs a year. They can live to be 5 to 7 years old.

Light Sussex

Light Sussex hybrids are a cross between Rhode Island red and the traditional light Sussex. They are a good choice for new chicken keepers as they have a lovely docile temperament. They enjoy being around people and are very friendly, they will follow you around the garden and always come running when you shake them tasty treats. I found that they are pretty good with children. You can look forward to some yummy fresh eggs as these girls will lay nearly every day. These girls will lay 260 to 280 medium to large light brown eggs a year. Light Sussex will become the grandmother of the flock as they live to the age of 8 years old.

So, these are some of the breeds that I use and I recommend to you. With all that in mind, we can finally move on to picking your girls.

So the time has come to pick your new girls

Before we go any further, my advice to you is pick the amount of chickens that you would like and stick to it. When selecting your hens you can pick them from different flocks and put them together. It is much easier to build a mixed flock at the start but it is much harder to introduce new chickens to an existing flock later on down the line, it can be done but it is stressful. I will touch on this a bit later but that is my advice to you.

When choosing your girls there are going to be a few things that you need to look out for, to make sure they are healthy and ready to go home with you.

First things first, don't feel pressured to buy immediately. Take your time and have a good look around. Inspect each chicken that you are interested in. I know it is tempting to go for the prettiest one straight away, however they might not be as healthy as you think.
So as you're looking, go through this list I'm about to provide you with in your head. Make sure all the boxes are ticked. This way you ensure that you go home with healthy and happy chickens and it will cause you less problems in the future.

You will need to pick up the girls that you are interested in so you can closely inspect them. The supplier should not mind you doing so.

How to pick up a chicken
Bend down and place both of your hands over both of the girls' wings and pick her up.
You can do this next step one of two ways, it's a personal preference to what you find more comfortable and easier to do.

You can hold the chicken against your body keeping one hand on one wing and its other wing firmly against your chest.

Or, you can firmly put the hen under one of your arms, securing one wing and the other wing against the side of your body.

If a wing gets loose and the chicken starts flapping, put her down and try again. Or else you'll have your arms outstretched holding some crazy flapping chicken as far from your face as possible, whilst everyone else is laughing at you.

Your checklist of things to look out for

Posture and walking
Pay attention to the chickens walking. They should walk around with their head up high and their tail facing upwards, I call it the chickwalk. As they walk, they should raise their legs purposefully and elegantly. If you see a tail tucked down for prolonged periods of time, this indicates that the chicken is unwell or something is wrong. If you see this I would avoid buying that particular chicken.

Eyes and nose
Look closely at the nostrils. I understand the nostrils are very thin and small but you must take a good look. They should be completely clear of any liquid, or mucus. They should be clean with no blockage and with no discharge.
The eyes should be clean and clear, they should not be watery or foamy.

Breathing
Put your ear to the chicken's chest. It shouldn't peck you, however I can't promise you that they won't. Listen carefully, it should sound nice and clear and you shouldn't really be able to hear them breathe.

If you can hear their chest rattling around almost like a mucusy sound then this is a sign of a respiratory problem. If you hear this again with a different girl then my advice to you is to not buy from that supplier.

The comb

The comb should be a pink to a reddish colour if they are at point of lay. If it's pale then it means you may have quite a few weeks before they come into lay.

The ladies legs

The girl's legs should be clean and smooth with no damage and no buildup underneath the scales. Check there is no damage to their feet and no hard buildup of dirt.

Feathers

When looking at the girls, take a look to see if they're pulling out feathers. The feather should look beautiful, smooth, glossy and clean unless they've just had a dust bath. Pull back the feathers and take a close look at their skin. Look for any parasites, red mite are grey before they start biting and sucking the blood so take a close look for them. Also, look for white eggs and lice. They are golden to brown. If you see any signs of the above I'd put the chicken down and move on.

The fluffy butt end

Finally, check the girls' butts. The technical word is vent, but I prefer butts. Their bottoms should be clean and the feathers should be clear of any faeces with a soft, fluffy appearance of the feathers.

With all this done and you've found your girls, it's finally time to take them home and introduce them to the family.

Time to take your hens home

So you're ready to take your girls home but how do you do it? It's pretty simple. All you need is a good strong box with holes for ventilation. If you have a pet carrier then you can use that as well. Make sure they are securely fastened within the vehicle. Drive as smoothly as possible and turn off the radio to reduce as much stress as possible. FYI you may want to crack a window, just trust me.

Now you're home. What do you do next?
The later in the day that you pick up your chickens the better. The reason for this is once they are home they can go straight into the coop and snuggle up for the night. This will minimise a lot of stress for your girlies.

If you are not able to do this, then put the girls straight into the run and let them explore their new home. Make sure the door to the coop is open so they can take a little peek inside.
I know it is tempting to fuss over them and invite the entire family around to show off your new addition to the family, however, my advice to you is to leave them alone and let them calm down and settle in. Make sure their food and water is available to them so they know where to find it in the morning.

Chickens are up from dawn till dusk. When the sun sets, You may have to tuck them into bed for the first couple of nights, as they may not know where to go.

You will need to let them out every morning at dawn. If you don't, they will be itching to get out for their breakfast and probably be shouting at you from the end of the garden until you come to open that door.

Remember for the first week, make sure you put poultry tonic in their water. Due to the stress, your girls would have lost a lot of vitamins and minerals, poultry tonic will replenish that and help them get over the stress.

Pecking order

I'm sure you've heard the saying 'pecking order'. Well this is where it comes from. Chickens will have a pecking order in every flock and they should establish that very quickly. However, this can take up to a week or two.
The best thing to do is leave them to work out the pecking order for themselves, unless one is really being a bully, then you have to intervene.
Water spray generally does the trick. If that doesn't work, I suggest isolating the bully, try a day at first. If that doesn't work, then try two or three days until she gets the idea and she will lose her place in the pecking order. This should help calm her down and help the others that have been bullied establish their place.
Because the girls are pecking at each other, you may find that blood has been drawn. You will need to clean it with antiseptic spray. Monitor the situation carefully because the chicken that is bleeding is more likely to be pecked even more by the other chickens. This could potentially cause even more bullying.
Establishing a pecking order is another stress added to the chickens which is another good reason to use poultry tonic.
This is the reason why you buy all of your chickens at the same time. You will find that the pecking order will be a lot less violent and they'll get along quite quickly and happily.

After a couple of days, start spending time with them. Let them get comfortable with you and the family. I found picking the chickens up and holding them a couple of times a day will help them become more attached to you. You want to do this before they start laying eggs. You will know when that is because when you go to pick them up they will flare out their wings and crouch down to protect their body. This is a sign that they are about to come into lay or laying and this makes it much more difficult to pick them up.

Shaking the tub of mixed corn will certainly get them used to you and understanding that you're the one that feeds them. They will love you for it, but remember only a little.

Now it's time to enjoy your girls and look forward to a nice supply of eggs.

Keeping your girls entertained

As silly as it might sound, it is important to keep your girls entertained, especially when they are spending a lot of time in the run. By keeping them entertained you are making them happy and happy hens means happy wife! no sorry wrong saying, I meant to say, happy hens means healthy eggs. I'm not suggesting you install a flat screen TV with a Playstation. But there are some simple things that you can do to keep them busy for hours and hours on end and you will probably find that you will waste the same amount of hours just watching them.

. Chickens are incredibly vain. They absolutely love to check out their comb and whip it around in the mirror. so they will love you if you get them a good quality plastic mirror.

. Dust bath, like most ladies, they will spend a lot of time in the bath.

. Hanging a cabbage is great fun for them, they love cabbage and they have to work for it.

. Chicken swing, now you're talking, who knew that chickens can swing and they love it. I will say it's still quite hard to get your head around watching a chicken just sitting there swinging, like it's a normal chicken thing to do.

. Xylophone, just to wind up the neighbours a bit more, but it's funny to watch and they do enjoy it.

. Omlet chicken peck toy, chickens will peck at it all day long. It is designed to release small amounts of mixed corn when pecked. A must have toy. Please only use it a few times a week to avoid overindulging on mixed corn.

Housekeeping

Housekeeping plays a big part in keeping chickens. It is important to have a strict cleaning routine. A clean environment will really help you in preventing a lot of problems for your girls.

By regularly keeping the coop and run clean you prevent the buildup of red mite and their eggs. It will also really help in preventing sickness and respiratory disease. The ultimate product for this is poultry shield.

Poultry shield

This is brilliant for cleaning the entire coop as it breaks down faeces easily and any bacteria. Poultry shield is also amazing at fighting off red mite,

Poultry Shield kills red mites by breaking down their wax coating, causing them to dehydrate and die. It also penetrates the protective coating of mite eggs, which destroys them and breaks the reproductive cycle. so it is very important to have this type of product in your armoury.

Daily clean

A daily clean is really to control the build-up of faeces and to check for red mite or any unwanted items in the house.

. All the butt nuggets should be removed from the COOP each morning to prevent the spread of parasites, bacteria and disease.

. All droppings should be removed from the RUN each day for the same reason.

. The water drinker should be washed of any debris or dirt and refilled with fresh water.

. Food should be topped up if needed.

Deep clean

A deep clean should be done every week to every three weeks depending on how many hens you have.

. You need to remove and dispose of the bedding in the coop and nest boxes.

. Spray the exterior and interior of the entire coop with Poultry shield. Allow that to soak in for 30 minutes.

. Respray the entire interior and exterior of the coop with Poultry shield, then scrub all of the surfaces. Make sure to get in the corners and cracks. This can be done with a scrubbing brush and a microfiber cloth.

. After all surfaces have been cleaned, washed down and dried, apply a small amount of diatomaceous earth around the coop, concentrating on the cracks and corners. There is an item called a powder puffer that my wife loves. It blows the earth easily around the coop.

. Now you can replace the bedding with clean, fresh bedding for your girls to snuggle up into.

Depending on how your run is set up, it is well worth using a pressure washer to wash down the ground and surfaces. I have soil and hardwood chips, so I remove the wood chip, then pressure wash the wood chips to remove all of the faeces buildup, then I can reuse it over and over, Which is a good way of saving money but keeping the run hygienic and the girls happy. Yes, it is a bit more work but, I don't mind.

Creatures of the night, red mite

RED MITE starts off grey 0.5 mm in size, then they feed on the host's blood and will turn red and grow to 1 mm in size. These horrible little things will lay hundreds of eggs every day. Red mite are something you never want, as it can make your girls sick, so prevention is so important. The art of preventing red mite is a strict cleaning routine and my favourite, Poultry shield, but, despite all your hard work, sometimes the vampires will come out to play.
There are a few ways that you can find out if you have a red mite infestation.

. The first sign you will notice is your girls looking more irritated, feather pecking and over grooming.

. If you have an infestation of red mite, it can make your girls very lethargic and they can become quite unwell. These are a few signs to look out for.

. You may also find that they are reluctant to go into the coop at night. This is another indication that you have red mite within the coop.

. Red mite can lead to a drop in egg production.

. The last thing to do is to put a small white towel in the corner of the coop at night. In the morning, If it is red then you most certainly have an issue.

If this is the case and you have an infestation of red mite, you will have to do a deep clean every 2 days.
You will have to change the bedding every 2 days and reapply the diatomaceous earth.

Old bedding should be disposed of in a solid sealed plastic bag and dumped at the far ends of the earth.

Then you'll have to spray the entire interior of the coop every 2 days with Poultry shield and let that break down the red mite and the eggs.

Make an extra effort to spray the nest boxes, underneath the perch, the ceiling of the coop and run and all the nooks and crannies. This process will have to be done for 14 days or until the red mite has completely gone.

Red mite on your girls

Removing red mite from chickens is a little bit more difficult. However, I have found some very good ways of doing this.

. Gently rub diatomaceous earth into the feathers of your hens. The aim is to cover them well but not to the point of a dusty old ornament on your grandmother's top shelf. Too much dust can cause a respiratory problem so don't go crazy. This can be done every 2 days.

. Make sure the girls have a dust bath available at all times. Use sharp sand, multi-Purpose soil and a little diatomaceous earth.

. Although this is not advised or licensed by the company, you could in extreme circumstances, use poultry shield. Spray this onto the girls avoiding their heads entirely. This will kill off the vampires just like holy water does. The choice is yours whether you choose to use it or not.

. Mite Treatment Large Birds Ivermectin 1% drops 5ml bottle, works wonders. The main ingredient is Ivermectin, this is the same ingredient that a vet will prescribe you, as it's used for the treatment of mites and lice. Myself as well as many other chicken keepers have used this product for their chickens and have had great success. One drop per 500g of body weight per week for three weeks; apply the appropriate dose once per week. The entire flock should be treated.

Do not eat your hens eggs during this treatment and for a further 10 days after treatment.

. You could also use Exzolt 10 mg/ml solution for use in drinking water for chickens. This works perfectly against red mite.

. The last thing you can do is consult a vet. They may be able to assist you further.

Lice

By pulling your girls feathers back you will be able to see if your girls have lice. They are about 3 mm long and look similar to a grain of rice.

They will look brown to golden in colour. Your hens will show similar symptoms to red mite if they have lice.

Lice are usually the cause of other animals visiting your coop at night, for example, foxes, rodents, etc.

If your girls have lice then you will do exactly the same as if you had red mite. However, the deep clean should be done 4 to 5 days apart for 14 days or until the lice have completely gone.

Again, Mite Treatment Large Birds Ivermectin 1% drops 5ml bottle, will work wonders for you.

As lice have a three week life cycle, each chicken will need to be treated. One drop per 500g of body weight per week for three weeks; apply the appropriate dose once per week. You will need to treat the entire flock. Do not eat your girl's eggs during treatment and a further 10 days after.

Worms

If your girls have worms in their faeces then it is important to make sure that all the faeces are picked up on a daily basis.

Once they are over the worms, you must remove all the bedding, then thoroughly clean the entire coop with Poultry Shield. Once it has dried you can put new bedding in the coop.

Then you must clean and sanitise the ground throughout the run, as they can become reinfected with worms by contaminated faeces.

You can do this by using washing up liquid or poultry shield, to spray and wash the ground.

Then you should use a pressure washer to thoroughly clean away any poop remaining.

The fluffy butt nuggets

This is the fascinating section that I mentioned earlier, poop! I bet you didn't realise you would be entering a world of poop. Poop, poo and more pooping. Each chicken will poo between 11 and 15 times a day. It's pootopia, sorry, I had to say it.

You will spend a lot of time looking at butt nuggets whether you like it or not. You will also notice many different types of poop.

Believe it or not, it is important to recognise the different types of droppings. By looking at your hens faeces you can tell a lot about their health.

Please keep in mind when you are looking at your girl's poop, not to panic every time you notice something different. The only time you should start panicking is if it becomes consistent or if you see it on a regular occasion. There are only a few things to panic about and you will definitely know once you get used to looking at poo on a regular basis.
My advice to you is, if you are unsure to seek professional advice, either the supplier or a vet.
As there are so many different types of poop, I'm going to go over the basics to give you an understanding of what to look out for. So here we go. Let's dive in.

Normal poop
Normal droppings are a mixture of poo and urine that can vary in colour, consistency, and shape. It will vary in colours of brown.

It should be well formed and compact, Almost tubular. Normal chicken poop has a white urate cap that covers about a third to half of the nugget, almost like a cherry on top of a cake.

As chickens poop a lot, you may also find that the poo will be looser now and then and that's ok too, as long as it's still reasonably formed.

Normal poop

Cecal poop

Cecal poop is normal, healthy girls will produce cecal droppings two to four times a day. These droppings are usually loose and soft looking, sticky, a little bit more smelly, and look like mustard, toffee, or chocolate sauce. Cecal poop is a good indicator that your girl's digestive system (tract) is working how it should.

Cecal poop

Intestinal lining poop

Intestinal lining poop will mostly happen at night. So when you pop down the end of the garden to clean out the girls, then to your horror you see red streaky tissue lining in and around the poop, don't jump for the panic button. This is just your girls shedding intestinal lining which is normal if it is now and then. If it's consistent, I would consult the supplier or the vet. Blood is a different story.

Intestinal lining poop

Bloody poop

Tiny specks of blood could be intestinal lining shedding. But a good amount of blood which is noticeable is different all together. Again, there can be a few factors to why blood is in your girl's poop, but the most common cause is internal parasites or Coccidiosis.

Coccidiosis is a gastrointestinal illness which is caused by parasites that make the gut lining bleed. You will notice that your girls will become lethargic, have diarrhoea, and have a lack of appetite. You'll probably find that the tail will stay tucked down. This can progress very quickly and your chickens can die. So it is important to act fast.

My advice to you would be to go and see a vet. They will test your hens faeces to determine if it is definitely coccidiosis. There are a few medications to treat coccidiosis. However, none of the medications have actually been approved in the UK to treat coccidiosis for LAYING HENS.

This means your vet has to prescribe medication under the rules of the Veterinary Medicines Directorate Cascade. Which means, in the UK, there are medications that will treat coccidiosis and they are safe to use for your hens, but as it has not been approved for laying hens, you can no longer sell your eggs to the public. However, you will still be able to eat the eggs yourself.

If the vet does prescribe medication for coccidiosis, it may not be specifically for chickens. It may be intended for pigeons or other species of birds.

You cannot eat the eggs during treatment or 28 days after treatment. You will no longer be able to sell your eggs to the public after the treatment of coccidiosis, as the medication has not been approved and there is not a 100% guarantee that it won't contaminate the eggs.

You could also use 'Harkers Coxoid oral solution' the ingredient is Amprolium. This ingredient is one of the medications a vet may prescribe for chickens with coccidiosis. This is licenced for pigeons. Amprolium has been approved in the United States, but it has not been approved in the UK for laying hens. However, myself as well as many other chicken keepers have used this product with great success. The internet will say differently as the information is intended for the United States but again, in the UK you cannot eat the eggs during treatment or 28 days after treatment. You can no longer sell your eggs to the public after this treatment. If you choose not to use it, then you must see a vet as soon as possible.

Bloody poop

Bloody poop

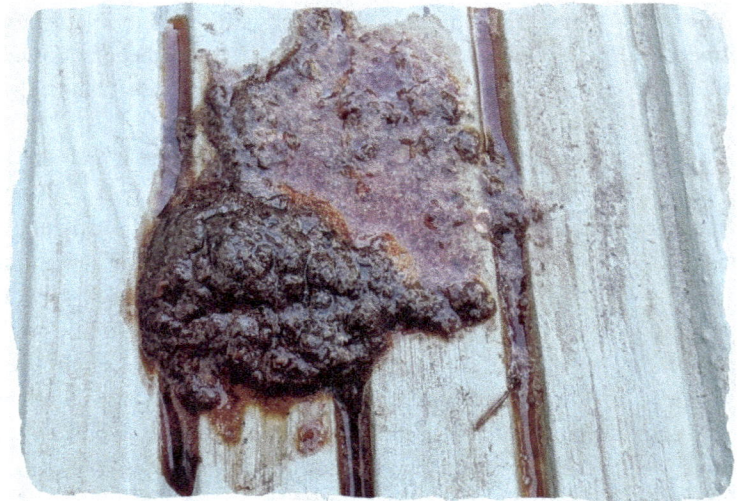

Diarrhoea

Diarrhoea is a difficult thing to diagnose. There are many factors to why chickens can have diarrhoea and it is important to understand why, so you can help her in the best way possible. It could be internal parasites or worms, so if your hens haven't been treated for worms in some time, then treating them for worms may help. If it's the summer then it may be heat stress, so make sure they have plenty of fresh water and shade. It could even be that they have too much protein in their diet or their food could be damp or mouldy. Try to support them by replenishing what they have lost, by adding Poultry Tonic to their water. If it is mild diarrhoea, there is also another product you can try called 'Chicken Squits'. If you have ruled out the above and there is no sign of the diarrhoea passing, then I would contact the supplier first, they may be able to provide you with antibiotics. If that fails, contact the vet for further support. If you see runny poop mixed with a white discharge, then that's more likely to be vent gleet, a delightful topic I will cover later.

Green poop

If your girl's are eating greens, plants or grass, you will often find that your hens poop will be green. If they eat grass you will see grass in the droppings, all this is normal. If your girls do not have access to greens or grass then something more sinister is at play. It's more likely to be a digestive issue. If the food is not being processed correctly, it causes bile and appears green in the poop, this can be caused by parasites, worms or a Impacted crop. It could also be that your hen is not eating nearly enough food. This leads to another problem, if a chicken is not eating then something is wrong. There will be an underlying illness somewhere or it may be an Impacted crop. If it's none of the above, then it may be something more serious, possibly Marek's disease, Avian influenza or Newcastle disease. If you have ruled out the above and your girl has green poop over 4 to 5 days and there is know sign of it going away, then it's best to go and see a vet.

Normal green grass poop

Fed no greens or grass, not normal

Worms in poo

Believe it or not, this one's pretty obvious. Worms can happen for many reasons. It can be stress induced. Or they could be feasting on slugs, snails, or earthworms that are infected with the parasite.

Wild bird faeces is also a cause of worms, as it carries a lot of disease. This is another reason why it's a good idea to have a roof over your coop and run.

There are several types of worms and they vary in colour and size.

If you are paying attention to your girl's poop, then you will certainly notice that your girls have worms.

You won't see spaghetti bolognese nicely presented on a plate, but you will notice thin worms in and around the poop.

Luckily this is reasonably simple to deal with if you have the right product.

Flubenvet layer pellets, this should be fed to your girls for 7 days straight. Do not feed your chickens anything other than Flubenvet during treatment.

My advice to you is, only buy this product when you need to worm your girls as it has a very short shelf life.

For a reference 5 kg will roughly feed 4 to 5 hens for a 7 day treatment.

Other ways of preventing worms in your chickens is to put crushed garlic in their water.

If the worms are persistent and not going away, which is unlikely, but it can happen, then the best advice I can give you is to see a vet. They will probably take a faeces sample and find out the cause then they will be able to help you further.

So now that's all cleared up, we can move on to health and diseases. As you can tell this book is full of joy.

Health and well-being

In an ideal world we would like to think we would have many years of blissful peace and quiet, without any problems and your girls happily pecking around the garden, laying you a lot of yummy eggs. And for the majority of the time this is the case.
However, there will come a time when one of your girls has a problem and it is best to be prepared for when that day comes.

You will find that most problems can be solved without having to visit the vet, but, if the problem persists, then as I've said before, my advice would be to book an appointment with the vet or contact the supplier just to be on the safe side.

Just a word of caution, when your girls do become unwell It is incredibly tempting to jump on your phone and search the internet for an answer, which my wife does love to do, for everything.
"It's on the internet so It must be true."
No wrong. Sometimes, yes, it is useful. However, most of the time it is very conflicting, false or misleading information that will not help you and more often make you worry even more than you really need to.

You will learn from your own experiences, this book or someone that you know that has experience with chickens.

This is why it is important to have a trustworthy supplier as they will be able to help navigate you through with their years of experience or a reliable vet that you can contact for advice.

Prolapse

Although there is not a high chance of your girls having a prolapse, there is still a chance, as your girls are laying eggs on a regular basis. A prolapse can happen for many different reasons. And in this case, prevention is better than cure. I'm not going to lie to you, it's quite a shock to see this bulbous thing hanging out of your girl's buttocks, staring at you like it's about to explode. As you can imagine, this is very distressing and uncomfortable for any chicken to go through, so it's definitely worth preventing it if you can.

Why a prolapse occurs

If your girls are laying eggs that are too large or eggs that are too soft, (we call them soft shell eggs), then this can cause your girls to strain much more than they need to, which will increase the chance of a prolapse.

If your chickens have a poor diet, especially if they have a lack of calcium, then this can lead to soft shell eggs. Or it can weaken the muscles around the cloaca increasing the chances of a prolapse.

A prolapse can happen as a result of an undetected abdominal or oviduct infection.

Age can also be a factor, as your girls get older and into retirement, their muscles don't work how they should and the muscles in the lining can stretch, increasing the chances of a prolapse. Just ask your grandmother how it feels and she will tell you everything you need to know.
But as daunting as this all sounds, there are ways that it can be prevented.

Preventing a prolapse.

. Make sure your chickens have a good quality layer pellet with the adequate amount of calcium that meets the nutritional needs for laying hens.

. Make sure that your girls have plenty of exercise. Pilates, spin class or circuit training, whatever they fancy really.

. Sunlight will definitely help as well. They love to sunbathe and it is also good for them as it is a natural source of vitamin D.

. If you notice that you are regularly getting soft shell eggs, then you will need to investigate and find out which one of your hens are laying the soft shell eggs. They will need to be correctly treated to prevent a prolapse. The most common cause of soft shell eggs is a calcium and vitamin D3 deficiency.

If a prolapse does occur then you have two options. You can visit the vet, or, you can roll up your sleeves, brace, and do it yourself.

The second you notice a prolapse you need to deal with it immediately.

How to treat a prolapse

Separate

As soon as you notice a prolapse you will need to separate the hen immediately from the rest of the flock. A small dark box or crate is ideal as you want her to move as little as possible And not be pecked by the other girls.

Wash infected area

You will need warm clean water in a washing up bowl, then add Hibiscrub, Savlon or Iodine. This will help disinfect the area of the prolapse.

Hold your girl with her bottom in the water. This will loosen any debris and help to clean the exposed tissue. It will also help to soften the prolapse ready for you to push it back in. Make sure the tissue is completely clean before going any further.

Pushing the prolapse back in

This is much easier to do with two people. So whoever's the most squeamish should hold the chicken. Put on clean rubber gloves and lightly lubricate.

You will then need to gently and slowly push the prolapsed tissue back inside the vent (bottom) and hold your finger inside the vent for 40 seconds.

Then you will slowly pull out your finger. That should work, if it doesn't and the prolapse comes back out, try again. If it continues to keep coming out you must go and see a vet immediately, as repeatedly trying to push a prolapse back in can cause your girl to strain, which may cause an even bigger prolapse.

Treating the tissue area

So now you'll need to treat the area with 'Preparation H'. This is a hemorrhoid cream that works wonderfully. Not that I would know personally!

You will need to treat the affected tissue internally and externally until your poor girl has fully recovered.

Preventing her from laying eggs

You need to prevent your girl from laying eggs for a while. This can be done by allowing her to rest and recover in a small dark crate. This will help prevent her from moving around too much. Avoid feeding layer pellets during this time and only feed mixed corn.

Water aid

While your girl is isolated it's best to add vitamins and minerals to the water that contains high amounts of calcium. This will help her in the recovery process.

Once she has fully recovered she can be allowed to go back in with the flock. Just be careful when you reintroduce her back into the flock as she may have lost her position in the pecking order, so you may see a bit of bullying. Monitor it carefully and deal with it accordingly.

Egg bound chicken

Egg bound chickens are rare, so I'm covering this just to be on the safe side, if it ever happens to you. If one of your girls is egg bound / binding it means that she is unable to pass the egg through her reproductive tract. This is a big problem for your girl, as it will not fix itself, meaning, it is extremely life-threatening.

My advice to you in this situation is to take your girl to the vet as quickly as possible. They are much better equipped to help her in this situation than you are. They will do an x-ray to see where the egg is positioned in the body and if it is still intact.

Then they will either perform surgery to remove the egg or use hormonal injections to encourage muscle contraction, this will help her pass the eggs. If it is not possible for you to get to the vet, then I will take you through some ways that you can help her. However, the odds are not in her favour and there is a possibility that you could lose her, if the eggs can not pass through her.

Causes of egg binding

. Soft shell eggs, if the egg is too soft, then it might not be hard enough and struggles to go through the reproductive tract, which could lead to egg binding.

. Eggs that are too big can get lodged in the oviduct.

. An unbalanced diet and poor nutrition may cause egg binding.

. Infection, especially in the oviduct, can cause swelling making it difficult for your girl to pass the egg.

. Stress can also cause your girl to become egg bound. If your girl has had a big fright or has become stressed because of changes to her environment or situation, then this may upset her egg laying pattern. Consequently progressing to becoming egg bound.

Signs that your girl is egg bound

. A very noticeable sign of your girl being egg bound is she will walk like a penguin.

. You will also notice her straining, as she tries to lay her egg in the nesting box.

. You will obviously notice she is not laying any eggs.

. You may also notice that she is reluctant to move, look weak and lethargic.

. Heavy breathing is another symptom that you may see, due to the pressure on her organs.

. A lack of appetite and reluctance to drink.

. You will also see a bulge in the abdomen close to her bottom.

How to treat an egg bound chicken

. If you can, run a warm bath with Epsom salts. If you do not have Epsom salts, then just run a warm bath. Place the bottom half of her body in the bath for 30 minutes, try to keep her calm during this time and gently massage her abdomen from front to back.

. Once she is out of the bath, gently dry her off with a towel and try to keep her in a nice warm place.

. Next, you will need Vaseline to lubricate her bottom. You will also need to gently lubricate just inside her bottom, don't go in too far or you may cause more problems.

. The last thing to do is gently massage the abdomen from front to back to help move the egg, so it makes it easier for her to push and pass the egg.

. Keep a close eye on her, make sure that the egg has not broken inside of her. You will notice if the egg has broken inside her by seeing parts of the egg or eggshell around her bottom. If you do see this then she must see a vet immediately
If you don't see a vet then this can lead to egg yolk peritonitis, which is just as serious as egg binding.

The maximum amount of time that you can care for her is 24 hours. After this, she must go and see a vet if she has not laid the egg.

How to avoid egg binding

. Soft shell eggs, the best way to avoid this happening is to make sure that your girls have the correct amount of calcium.
Calcium will keep the eggs nice and hard and it will also strengthen the muscles, so the muscles will continue to contract and pass the egg through the reproductive system.

. A good quality well-balanced diet and plenty of fresh water, this will help reduce the chances of egg binding.

. Make sure your girls have plenty of exercise.

. Make sure their environment is as stress-free as possible.

Vent gleet

Vent gleet is quite frankly disgusting. Ok let's start again, so vent gleet isn't the prettiest thing to deal with, but there may come a time when you do have to deal with it. So, I will cover this lovely topic just in case you do. Vent gleet is a fungal infection that affects the vent (bottom). The most obvious sign of vent gleet is a smelly, yellowish-white discharge coming from your girl's bottom, which sticks to the feathers around the backside. I told you it's not pretty. The vent area may also appear red and inflamed.

In severe cases of vent gleet, your girls abdomen may be firm to touch, their bottom may be very swollen, and their poop could even contain blood. The good news and yes, there is some good news, it can be easily treated.

Causes of vent gleet

. Fungal infection, including yeast, is generally the main cause of vent gleet.

. PH imbalance. If your girl's body is too alkaline or acidic, it can make your girl more receptive to vent gleet.

. Bacterial infection.

. Parasites. Internal parasites can irritate the cloaca (bottom) and cause vent gleet.

. Stress and hormonal cycles impact the entire body and affects the digestive system. In chickens, those factors can lead to vent gleet.

. Damp or mouldy food can contribute to vent gleet.

Treating vent gleet

. Isolate your girl from the rest of the flock, ideally in the isolation pen. Make sure she has food and plenty of fresh water.

. You will need warm clean water in a washing up bowl, then add Hibiscrub. Hibiscrub contains Chlorhexidine, which is an antibacterial and antiseptic that can be used to reduce and prevent bacterial growth. If you don't have Hibiscrub then you can use Epsom salts, a lot of people do but it's not as good as Hibiscrub.

. Hold your girl's bottom in the water and let her bottom soak for at least 10 minutes. Put on a glove and gently clean the bottom, the skin and the feathers around the bottom.

. Take her out of the bath and gently dry her off with a clean towel. Then, trim the feathers around the bottom to keep the area clean.

. Apply Canesten cream around her bottom and just inside the bottom. You should reapply the cream 2 to 3 times a day.

. Add a little apple cider vinegar with garlic to her drinking water for 2 days. This will acidify her digestive tract and crop, which will help her in the recovery process.

. You will need to bathe her every 24 hours in hibiscrub and re-apply the Canesten cream until she has fully recovered.

If your girl is not improving after 7 to 10 days, then it may be best to book an appointment with the vet. They may be able to help you further, by investigating the severity of the infection and then providing a treatment plan to suit her situation.
Other than that, just make sure you provide your girls with a well-balanced diet, plenty of fresh water and a clean and hygienic living environment.

Respiratory disease

Respiratory disease is somewhat common in chickens. It can be caused by a number of things like viruses, bacteria, fungi and mycoplasma. It can even be caused by a dusty environment. Like a cold, respiratory disease is contagious to the other members of your flock. It is important to notice the signs early, as early prevention will make a big difference.
Sometimes it can just be something obstructing their nostril like a grass seed or a bit of dirt. You may even be able to see the object blocking the nostril, this will replicate a respiratory disease, causing their nose to run and them to sneeze. It can take quite a while to dislodge from the nostril. Regularly wiping the nostrils with warm water and a cotton pad will help dislodge the obstruction.

If you notice the symptoms of respiratory disease, you have a few options. You could contact your poultry supplier, they should be able to supply you with medicine that can be added to the water. This would be my first point of call.
Or if the symptoms are mild you could try first, Aviform Respiratory Stuff! It works well. In more serious cases, Denagard works very well, as it's designed to treat Mycoplasma in chickens. Both of these products can be purchased online. The entire flock will have to be treated to avoid the spread of the disease.
Most of the time this will work and you can go back to relaxing on the sofa without worrying about your girls.
But If these options don't work then a trip to the vet will be the next step. They will be able to assist you with a different treatment plan.
Wouldn't it be nice If they could just pop a few lemsips, tuck themselves into bed and be done with it. But of course that would be too easy.

Signs of respiratory disease

. Normally the first thing you will notice is discharge coming from the nostrils, a clear mucus, like a runny nose. If they've had a runny nose for a little while, you may find that the nostrils become blocked with hard discharge. This will need to be unblocked with warm water and a cotton bud or cotton pad. I personally think cotton buds work better.

. You may notice them sneezing, coughing or have difficulty breathing.

. When chickens' airways fill up with mucus, they tend to shake their head a lot more than usual to try and clear their nostrils. This behaviour becomes more noticeable as it becomes more consistent.

. You may also notice a drop in egg production. This tends to happen when they are unwell.

. A watery discharge coming from the eyes or even conjunctivitis.

. The last thing to do is pick up your chicken, by now you should pick your girls up like a pro. Place your ear to their chest. Listen carefully, it should sound nice and clear but If you can hear a raspy rattling sound, then it's a respiratory disease.

Preventing respiratory disease

There are a few things that you can do. Most of the time if your hens are healthy and happy, you are much less likely to have a respiratory problem in your flock.

. Making sure when purchasing your chickens, they are vaccinated against infectious bronchitis (IB) This will most certainly help.

. New chickens should be quarantined for 2 to 3 weeks before they meet your girls.

. Make sure their house is well ventilated and clean.

. Try to make the chicken's environment as dust free as possible.

. Make sure that they live as stress-free as possible, stress can sometimes lead to respiratory problems.

. Avoid things like oil-based paints and Aerosol sprays.

. Cigarettes are very poisonous for chickens and the smoke can cause a respiratory problem.

Impacted crop

I will explain the basic anatomy of a chicken, just so you understand how the crop works. The crop is essentially a bag, where food is stored after the chicken has eaten. It is located on the front right side of the breast. When the crop is full you will be able to see a small bulge on that front right side. Next, food will pass through to the stomach which is made up of two parts. The proventriculus where food is also stored. Then it will move down to the second part of the stomach called the gizzard. The gizzard will store grits that your chickens have eaten and that will act as teeth and break down the food. The crop will only allow a certain amount of food through at a time, to allow the stomach to do its job.

Impacted crop is where the chicken has overindulged to such a point, that the crop is completely full and has become blocked with food. Stopping the food moving down to the stomach.
This can be extremely uncomfortable for your girls.
If it's not treated It can develop to what's called a sour crop and this will need immediate attention. So the best thing to do is to prevent it rather than treat it.

How to prevent a Impacted crop

. Avoid using straw or hay bedding as they like to eat it and this can cause an Impacted crop. Unlike humans, we know we're eating junk food, they don't.

. Don't allow them to gorge on grass until the afternoon, let their stomachs fill up with layer pellets before allowing them to roam free and go wild at the buffet.

. Allow them to have access to fresh water and food all day long.

. Make sure their food is not damp or mouldy.

. Mix small amounts of chicken grit to their feed to help break down their food in the gizzard.

Signs of Impacted crop

. One of the first signs you will notice is that your hen is not eating.

. Your hen will more than likely stop pooping.

. The crop is big and noticeable in the morning before food.

. If you feel that your chicken has an Impacted crop, then the first thing to do is take away their food and water once they have gone to bed. In the morning before any food or water, examine the crop, if it feels big and hard then it's a Impacted crop.

. In the morning compare your chickens crop to the rest of your flock before food or water. If it is bigger and harder than the rest of your girls it's a Impacted crop.

If your hen is showing the symptoms of an Impacted crop, then it is time to deal with it and the sooner you deal with it the better to prevent a sour crop.

Treating a Impacted crop
. Separate your girl from the flock and only give her fresh water. Do not feed her for the first day.

. Massage the crop from the top to the bottom, this will help break down the blockage.

. Something that really works is olive oil. You will need to syringe a small amount, slowly into the side of her beak and down her throat.

. You will need to repeat this process at least four times a day but if you can do it more, great! It all helps.

. After 24 hours you will need to start feeding her again. You need to feed her food that is easy to digest.
Boiled egg, plain yoghurt, mashed fruit, diced or minced vegetables are the best for this.

For the next couple days continue with freshwater and digestible foods. You will need to continue with the olive oil and the massages. You will know if her condition has improved, when she starts to poop and her crop has shrunk to the normal size. If this is not the case and her condition has not improved then this will develop to a sour crop, meaning more serious intervention will be needed.

A sour crop is a fungal infection that's caused by food sitting in the crop for long periods of time, that then ferments. You will know if she has a sour crop because she will have a horrible sour smelling breath, which smells much worse than your partner's morning breath. Treating a sour crop feels a little bit barbaric. However, you need to be brave, roll up your sleeves and just do it.

Treating a Sour crop

. Syringe a small amount of olive oil down her throat and wait for 5 minutes.

. This can be done with one or two people, however it is easier to do with two people. You will need to hold her and tip her down. Almost like she's upside down.

. Whilst you are tipping her down. You must massage her crop, be reasonably firm.

. This must only be done for 30 seconds and then tip her back up so she can breathe.

. You will need to keep doing this until all the fluid has come out of her mouth. If food starts to come out you must stop as she can choke. You are only trying to get the fluid out, not the food. Do not repeat this process any more than eight times a day. This process should only be done for 2 days.

. Add a little apple cider vinegar to her water to help balance the crops PH and prevent harmful bacteria.

. If the poor girl is still not improving, then you must go and see a vet. They may do a surgery to cut open the crop and clear the blockage.

Bald bottom chickens

Moulting is a completely normal thing that happens every year, despite them looking pre-plucked ready for Christmas dinner. Please don't panic when you start seeing your girls pulling out their own feathers. Moulting normally starts from the head and neck and works its way down.
The first moult your girls experience will be between 16 to 18 months of age. I hope your bedside manner is good because your job will be to nurse them through the process.
The reason why chickens moult is to shed their old feathers to grow Nice shiny new ones. It also gives them a good needed break from laying eggs and your girls can build up their body reserves of nutrients. So you will find a massive drop in egg production and that is normal. Moulting should take 7 to 8 weeks from start to finish but sometimes it can even take longer.

How to care for your girls when they are moulting

During the moulting period your girls will lose a lot of vitamins and minerals, so this is where poultry tonic comes into play. It is great at helping them replenish what they have lost. Use poultry tonic with fresh clean water until the feathers are completely back.
Your hens will need a lot of protein to grow new feathers, as 85% of feathers are built from protein. So you will need to increase their protein intake with a higher protein feed, you can use chick crumbs feed, or broiler feed. You can also give your girls scrambled eggs with the shells to help boost the amount of protein they are getting, and they will love you for it.
Once they have finished moulting they can gradually go back on to complete layer pellets.

Please try to make the chicken's environment as stress-free as possible. They are going to feel a bit sorry for themselves as you can imagine. So you can help by providing plenty of space for them to rest and relax. Running around naked for most of the day can be a bit degrading for some people. So If you can, try and give your girls places where they can go and hide.

Please avoid handling them as much as possible. It's very uncomfortable and very tender during moulting, so handling them will be very uncomfortable and sore for them. You can cuddle them as much as you like once their feathers have grown back. Other than the points I've just gone through, there is nothing else you can really do to make this experience more comfortable for them.

Scabby Scaly legs

Chickens with scaly legs are caused by microscopic mites that burrow into your girl's legs and eat at them. Sounds charming doesn't it? Well, I can assure you it's very uncomfortable for your girls. The best way to deal with this problem is to dip your girl's legs in hibiscrub or surgical spirits for 60 seconds as often as you can, preferably once everyday.

This will need to be done until the girl's legs are better. That could be as long as several weeks, after you have bathed the girl's legs, you will need to cover the entire legs with Vaseline to suffocate the mites and it will help soften up the dry scales. This will also need to be done until they are better. If you have a bad case of scaly leg, then you can also use Mite Treatment Large Bird Ivermectin 1% drops 5ml bottle, to kill the leg mites. If you use Ivermectin do not eat your hens eggs during treatment and for a further 10 days after treatment. Other than that, there is not much else you can do. This is an effective method of treatment but it just takes time.

Eye issues

Eye problems are just one of those things that happen. As you can imagine, your girls spend a lot of their time with her face stuck in the ground, while their sisters scratch at another part of the ground flicking mud at their face, without a care in the world.
You can see why every now and then something will get stuck in there.
If you notice that this has happened to one of your girl's, then it's a few things you can try before you need an appointment with the vet.
Fill a syringe with clean water, flush out the eye gently and slowly until all the debris has gone.
If you notice the eye is irritating her or is scratched, you will need to ask your vet for an eye ointment, it will probably be an antibiotic gel, you will need to then apply this until the eye has fully healed.

Now, if you notice that your girls have bubbly clear fluid, foamy discharge or yellow pus coming from the eyes, then this is a sign of Mycoplasmosis.
Mycoplasmosis is a contagious respiratory disease caused by bacteria-like pathogens. However, you may also notice other signs like sneezing, coughing nasal discharge and gasping, in other words a respiratory infection.
Mycoplasmosis can be transmitted very easily. There is no cure for Mycoplasmosis, but the vet may be able to prescribe antibiotics to reduce symptoms and speed up your girl's recovery.
Or you can use Denagard, which you can purchase online. This is used for the treatment of mycoplasmosis in chickens and it can be an effective treatment.
If you are unsure as always, my advice would be to talk to the supplier or visit a vet.

Your girls and other animals

As you well know, chickens live on a farm with many other animals and they normally live very happily together.
This is also the case at home. My cats get along fantastic with our chickens and quite often the cats will lay on the lawn while the chickens peck around them, happily living together in harmony.
But please don't expect them to get along immediately.
Take your time when introducing your current pets to your chickens. Make sure any interaction between your current pets and your new chickens are supervised.
Please don't take any risks, chickens don't make great chew toys and it'll probably make the kids cry to see it.
If you do this part properly then you will be comfortable in the future leaving them with each other for short periods of time.
Everyone thinks that foxes are the chickens biggest threat and they are a serious problem, but in fact it is actually dogs, either the owner's dog or the next door neighbour's dog, so please bear that in mind if you have a dog.
Gradually and slowly introduce them, you never know they may become best friends or the dog becomes their personal bodyguard.
It does happen and it's pretty cool to see.
A word of caution, do not leave your chickens unattended in the garden if you go out for the day. If the fox decides to pay a little visit to the grocery store, your girls might not be there when you get back.

So as I've said to you before, as daunting as all of this sounds, It doesn't necessarily mean it's going to happen to you. Most of the time, keeping chickens is an enjoyable process with very little issues at all. However, fail to prepare, prepare to fail. It is best to have the correct information to deal with the situation at your fingertips, If that day ever comes. Reading and storing this knowledge will make you a better all-rounded chicken keeper as you will know what to do in different situations.

Introducing new chickens

Introducing new chickens to your existing flock can sometimes be a difficult thing to do, especially when you're not sure how to do it. People will give you different advice on how to do this. However, there is definitely a wrong way to do it and there is most certainly a correct way of doing it. When you have a smaller flock, everything is more intensified, so it's important to do it right. I am going to take you through the easiest and most effective way of introducing new chickens to your existing flock.

As you know already, taking a chicken from the farm and moving her to a new home is extremely stressful. So throwing them into a new flock as well as settling into a new home is overwhelming and simply too much. If you do it this way, you will more than likely see extreme fighting to establish the pecking order. The new girls will become isolated and extremely depressed. What will also happen is the new chickens will not go near the food, as your existing girls will peck them when they go to eat. This is the wrong way to do it and you will cause yourself unnecessary complications and problems.

The right way to integrate chickens in three simple steps.

. *Step 1.* This is best done with two chickens minimum, as they'll become best friends and protect each other from your existing flock. The first thing to do when you get your new girls home is to put them in the isolation pen, separate from the rest of the flock. This way they can get used to their new environment safely and as stress-free as possible. Most importantly, they can all get used to each other through the mesh, without pecking each other to death. The isolation pen should be equipped with a food dispenser and water drinker.

They will need to stay in the isolation pen for 2 to 3 weeks before you open the door and they can live with each other full time.

. *Step 2.* After the first week you can start introducing your new girls to your current flock but only for a few hours a day. This will need to be done in a large open space, like the garden. The reason for this is, if any of the new girls are being bullied, there is enough space for them to run away and stay safe. The second reason is, they can get used to eating with each other, pecking the grass or sharing the food dispensers. Please make sure there are multiple places they can eat their food or drink some water.
This is a great way of building them up to become best friends.

. *Step 3.* It's time for you to open the isolation door and allow all your girls to live together. But just before you do, make sure there are plenty of perches for them to jump up on and escape any pecking.
It is also very important to make sure that there are at least two food dispensers and two water drinkers. This will prevent your old girls dominating the food and water from the new ones.
Please keep a close eye on them to make sure that they are all eating.
There is always going to be a bit of pecking to establish the order. However, if you notice that one of your girls is being a horrible bully then the best thing to do is to put the bully into the isolation pen, not the chicken that is being bullied.
Try a day to start off with, if she continues to bully then put her in the isolation pen for 2 to 3 days. This will help the new girls establish their order and calm down the bully.
However, if the first two steps were followed properly, you will dramatically reduce the chances of this happening resulting in a happy, feathery family.

Producing the goods, eggs

What better way to end the book, than the bottom end goods we have all been waiting for, golden butt nuggets.

Eggs are one of the main reasons people keep chickens and for me it's definitely the fun part. You can't beat waking up in the morning, grabbing a coffee and walking down the garden to find a ton of lovely fresh eggs just sitting there waiting to be eaten. If your girls are producing the goods, it normally means they are happy and healthy.

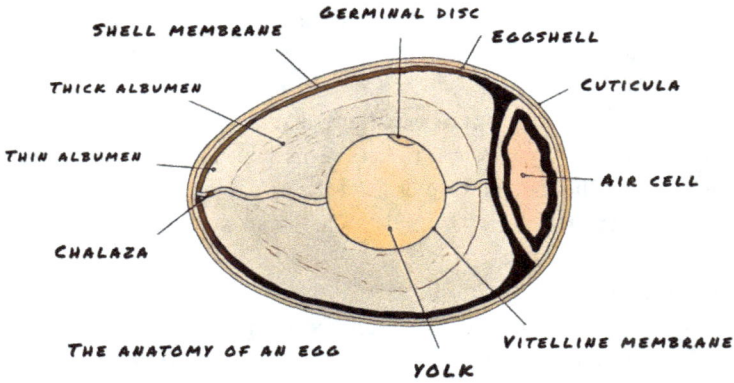

A chicken's main job in life is to lay eggs. They live for it. They love it, and they do it almost everyday, which means they are the elite laying machines.

Especially hybrid chickens, Which can shoot out over 300 eggs a year, Incredible really. However they can only do this with your help.

If you look after your girls, then I promise you the quality of eggs that they will produce is just amazing. You can't even compare them to supermarket eggs in my opinion.

Here is a little experiment for you and the kids. Try to pull out the yolk of a supermarket egg and firmly squeeze it, It will more than likely pop with ease.

But if you try the same experiment with your eggs, then you'll notice that you need to apply much more pressure to break the yolk. It's oddly satisfying and somewhat addictive so don't get too carried away or you'll have no eggs to eat.

This is one way you can tell, the eggs that your girls are producing are of great quality.

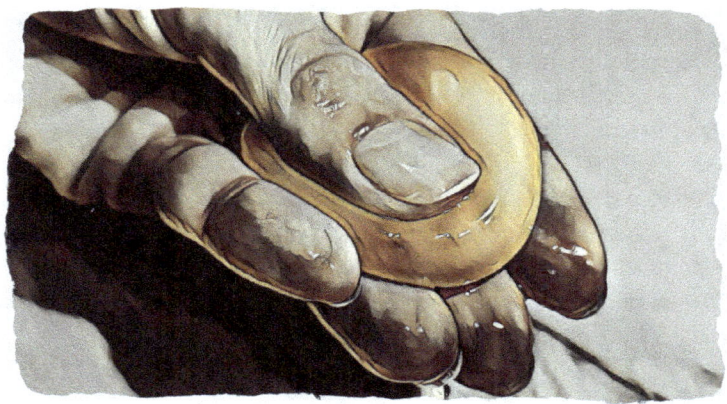

Maintaining a regular supply of healthy eggs.

. Providing a well balanced diet. A complete good quality layer feed should make up at least 90% of your girls diet and only 10% treats. The feed should include a minimum of 16g of protein, calcium for strong eggshells, amino acids, vitamins, and minerals for egg quality and your girl's health.

. Make sure your girls always have access to clean fresh water.

. Provide synthetic light. Your girls need at least 15 hours of light a day. When the light is less than 11 hours per day, your girls' egg production may slow down. You can provide artificial light with an incandescent 40-watt light bulb or LED light 13-watt per 90 sq ft run space. By doing this you are replicating daytime, so they will stay up longer and continue to eat for longer. This is what commercial farmers do to ensure a regular supply of eggs.

. Keep the nest boxes and coop clean.

. Keep your girls warm in the winter. You should provide your girls with a warm coop that's protected from the elements to encourage egg production.

. It also depends on what breeds that you choose to keep. This is why I personally keep hybrid chickens and that is why I have recommended them to you. They will keep popping out your breakfast on a regular basis, even in the winter, when the temperature drops and the nights close in.

As much as you love eggs there are only so many that you can eat. So what do you do with the rest of the eggs? Well I'm about to tell you how you can keep chickens and massively keep the costs down.

What to do with your yummy eggs

The answer to this is very simple, sell them. But just remember, don't forget to tell the tax man, or he'll be chasing you down the road for your last penny.

But seriously sell them. You're producing good healthy eggs and when your friends and family taste the difference, they'll be lining up at the door. Ask everyone you know to keep their old egg cartons, that way they can safely be transported to your customers. If you want to look a little bit more professional, you can always go online and purchase labels to go on your egg boxes.

You and Rodney won't be millionaires by this time next year but by selling your eggs, you can cover the costs of your girls food, bedding and other goods that you may need.

Now your egg industry is in full production. You want to make sure that everything is running smoothly. However, every now and then problems do occur with the eggs. So I will take you through some of the problems that you may face, to make sure that your business doesn't feel the effects.

Not laying eggs

If your girls get enough exercise, a nice healthy, well balanced diet and plenty of water. Then there is no reason why they shouldn't lay you lots of lovely yummy eggs.

If your girls are not laying eggs it could be a sign of illness, so the first thing to do is to make sure that they are happy and healthy. It may also be that they've been frightened by a predator, if this is the case, then it can take quite a while for them to come back into lay. Try giving them poultry tonic to restore the vitamins and minerals they would have lost in the fright. If all the girls are happy and healthy and have not been frightened by predators, then it could be a simple case of overindulging.

The best way to fix this is to stop giving them any treats, despite how tempting it might be. Just feed them layer pellets and fresh water until they come back into lay.

Soft shell eggs

I know we have touched on soft shell eggs but we need to dive into the subject a little bit further. If you notice that one of your girls is laying soft shell eggs, then the best thing to do is act as quickly as possible to stop her having further problems, like a shell gland infection or egg peritonitis. Most of the time it's a vitamin D3 and calcium deficiency or lack of calcium. This can easily be remedied by adding Aviform shell stuff! (Liquid) to their water, follow the instructions on the bottle. If this doesn't sort out the problem, then it could be something that's going to be more of a problem.

It may be a shell gland infection or infectious bronchitis (IB), if it's either of these, you will notice abnormal shaped eggs, soft shell eggs or possibly blood on the egg. If this is the case, you will need to contact the supplier or the vet to assist you further with a course of antibiotics.

Eating the goods

Now this is something that really ruffled my feathers. All your hard work has been eaten, and what's worse the girls don't even look guilty. You cant even torture them into an answer because they don't speak.

So you will need to deal with this before they get the taste for scrambled egg and your breakfast.

Preventing egg eating

. Collect the eggs as soon as possible, ideally in the morning before they have a chance to peck at them.

. Keep the nest boxes as dark as possible as chickens cannot see in the dark. What they can't see they can't eat.

. Sometimes chickens will eat the egg because they need more protein or eat the egg shell If they are lacking in calcium, so make sure their diet covers all of their needs.

If they are eating through your enterprise
. You can try fake eggs in the nest box. It can work quite well as once they keep pecking the fake egg and it doesn't break, they start to get bored and give up pecking the egg's entirely.

. Build or use a roll-away nesting box, This type of nest box has a slanted surface that causes eggs to roll away into a trough that the hen can't reach.

. Fill up a few of the eggs with English mustard, and put them in the nest box for them to peck at. That will teach them, ha!

So that brings us to the end of the book. Now it is time for you and the family to have many enjoyable years with your girls.

Just remember for the majority of the time, keeping chickens is easy and full of fun.

It is only now and then that you will come across the problems in this book.

However I wrote this book so you are prepared and ready to deal with an issue when it occurs. Of course this hasn't covered absolutely everything, but it does cover the majority of the problems that you will potentially face when keeping chickens.

I wish you good luck and many enjoyable years with your new additions to the family.

I would just like to say a huge thank you for purchasing my book. I really hope it has helped you and will continue to help you through your journey of keeping chickens in your garden. If you have enjoyed my book I would greatly appreciate your kind support by leaving this book a review on Amazon.

Available on AMAZON
By A M STOKER

Notebook: Your story will come to life once the ink touches the page

Shopping list: Complete with tick boxes

Roman and Bonnie's adventures

To purchase your copy on Amazon, you can search for the titles above or
you can search for the author
A M STOKER

www.ingramcontent.com/pod-product-compliance
Lightning Source LLC
Chambersburg PA
CBHW071056240526
45469CB00006BD/2322